629.4
Sha

Shapp, Martha
3853
Let's find out
about space
travel

3853

DATE		
OC 21 '72	OC 27 '80	
DE 1 '72	MAR. 16 1982	
MR 17 '73	JUL 8 1986	
MG 3 '73	SEP 03 1987	
JY 5 '74	JY 16 '88	
JA 6 '75	MR 04 '91	
JUL 7 1977		
APR. 20 1979		
AUG. 15 1979		
OCT. 4 1979		

© THE BAKER & TAYLOR CO.

LET'S FIND OUT ABOUT
SPACE TRAVEL

LET'S FIND OUT ABOUT

FRANKLIN WATTS, INC.
845 Third Avenue, New York, New York 10022

SPACE TRAVEL

BY MARTHA AND CHARLES SHAPP

Illustrated by Joel Snyder

10 9 8 7 6 5 4 3 2
SBN: 531-00068-0
Library of Congress Catalog Card Number: 70-131141
©Copyright 1971 by Franklin Watts, Inc.
Printed in the United States of America

LET'S FIND OUT ABOUT SPACE TRAVEL

Long, long ago, people looked up at the sky and
    wondered.
What is the moon made of?
How far away are the stars?
Is there any life on the planets?

Even by using a telescope, scientists could not
find answers to all the questions.
If only man could travel across space to land on
one of the far-off places!

But how?

One man thought that a balloon with wings could
travel out into space.

Another idea was that a big bullet with men in it
could be shot up to the moon out of a big
cannon.
Of course, these ideas could not work.

A long time ago, the Chinese people invented
rockets to use as fireworks.
Did you ever see a rocket shoot up into the sky
on the Fourth of July?
Now scientists find that they can make big
rockets that can travel way, way up and out
into space.

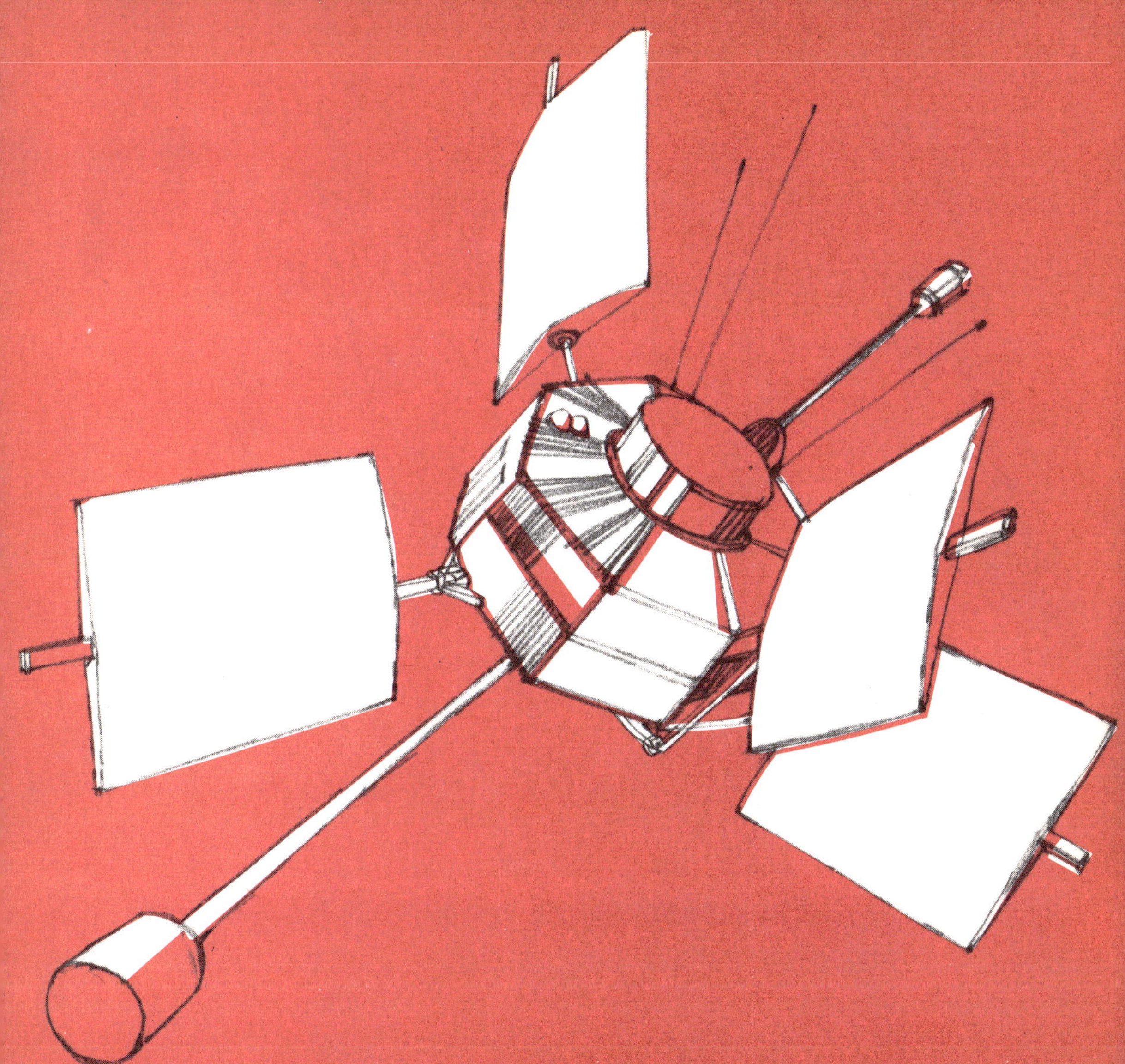

At first, rockets lifted only instruments into space.
Many such instruments are now in orbit around
the earth.

Some of these instruments send back messages
about the weather.
These instruments help weathermen tell us what
the weather will be tomorrow.

Some of these instruments help send TV pictures
across long distances.

Scientists were not sure that men could travel
  safely through space.
So they sent animals up in spacecraft first.
The animals came back unhurt.

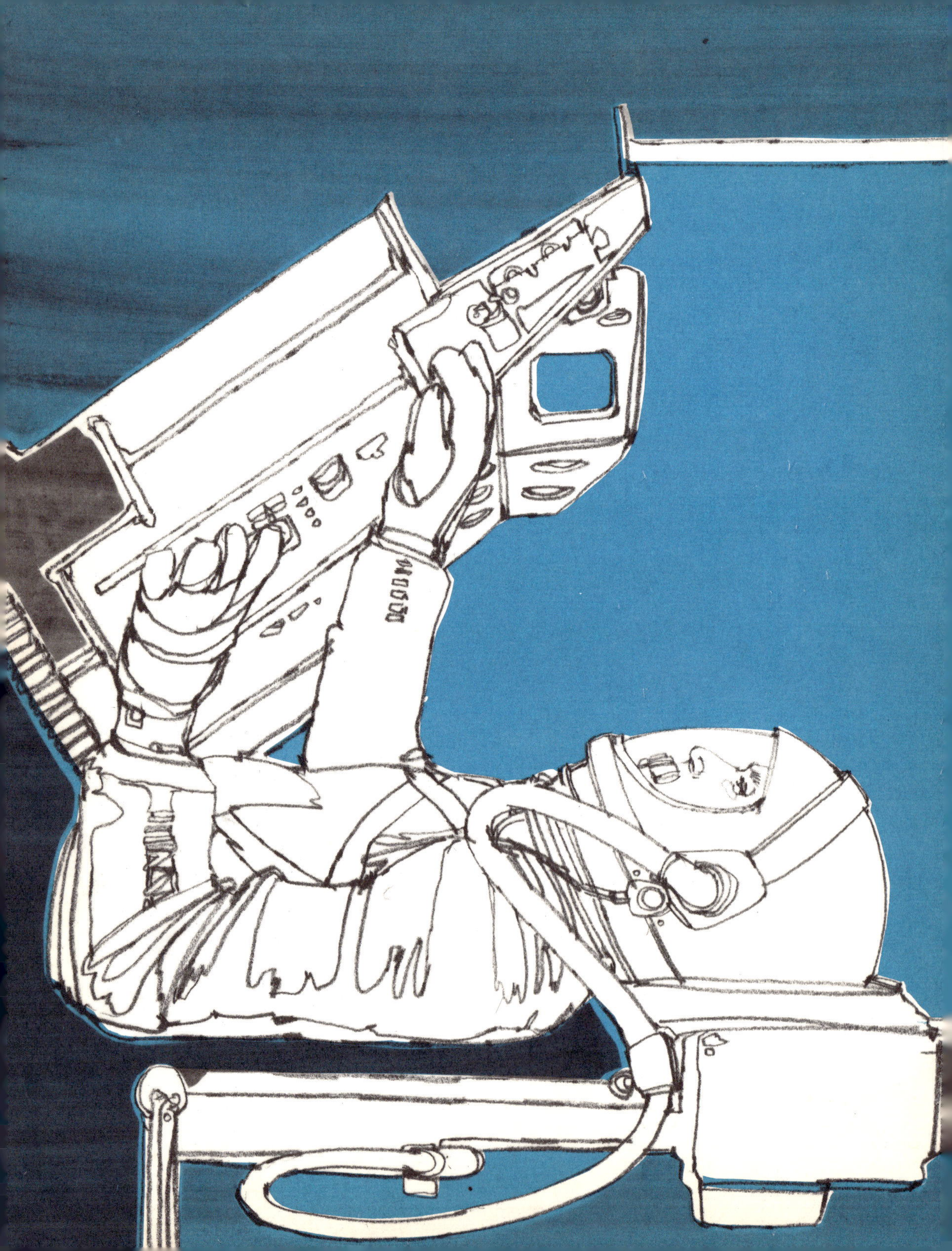

All was ready for man to travel in space.
The first manned flight was made by one
   astronaut.
Later, spacecraft with two or three astronauts
   were sent off into space.

The first manned flights were in orbit around the
earth.

Longer and longer flights were made.
Then three astronauts blasted off into space on a
trip that took them around the moon and back.

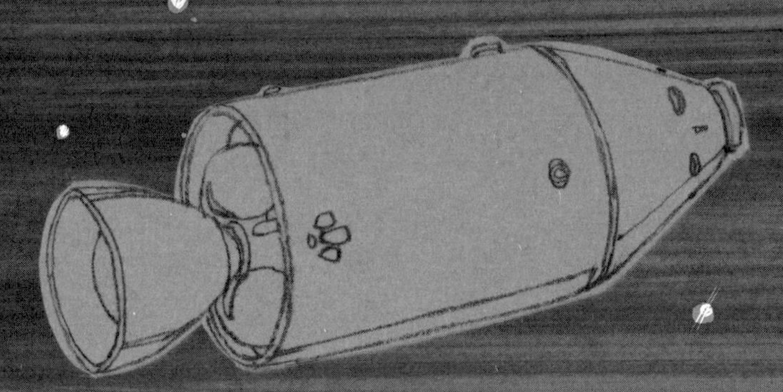

On July 16, 1969, three astronauts, Neil A. Armstrong, Edwin E. Aldrin, Jr., and Michael Collins, blasted off from Cape Kennedy.

On July 20, 1969, the whole world cheered as Armstrong and Aldrin set foot on the moon. Collins, in the spacecraft, orbited the moon awaiting their return.

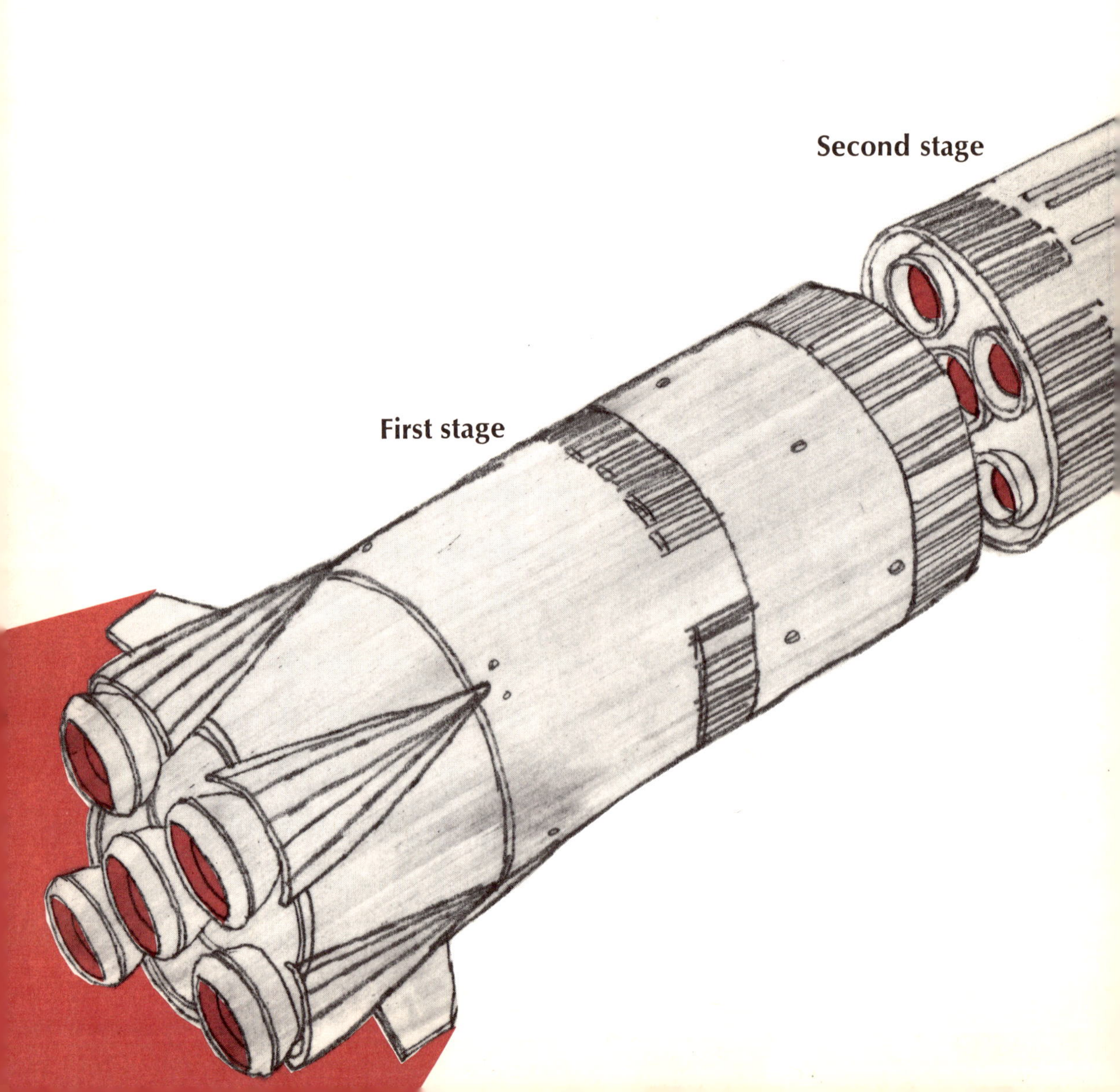

Second stage
First stage

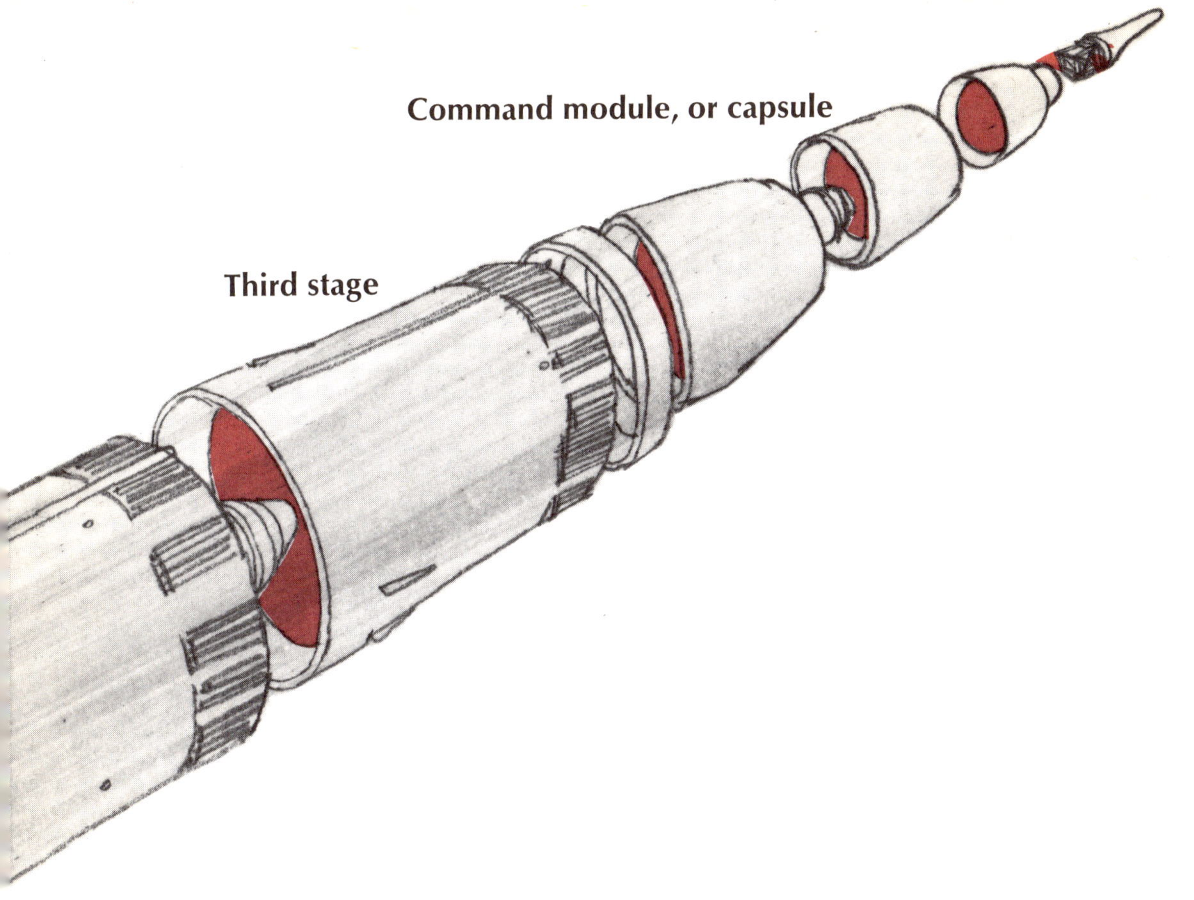

Space travel begins on a launching pad on earth.
The three rocket stages are full of fuel.
The astronauts travel in the command module,
  sometimes called the capsule.
The landing module is inside the service module.

The astronauts go up to the tip of the waiting
rocket in an elevator.
They are shut into the capsule while the rocket
is still on the ground.

Engineers in the launch control center watch
their instruments to see that all is well.
They begin the countdown: 9—8—7—6—5—4—
3—2—1!

Huge flames and clouds of smoke burst from the
   first rocket stage with a great roar.
Lift-off!
The tall craft rises slowly from the pad.

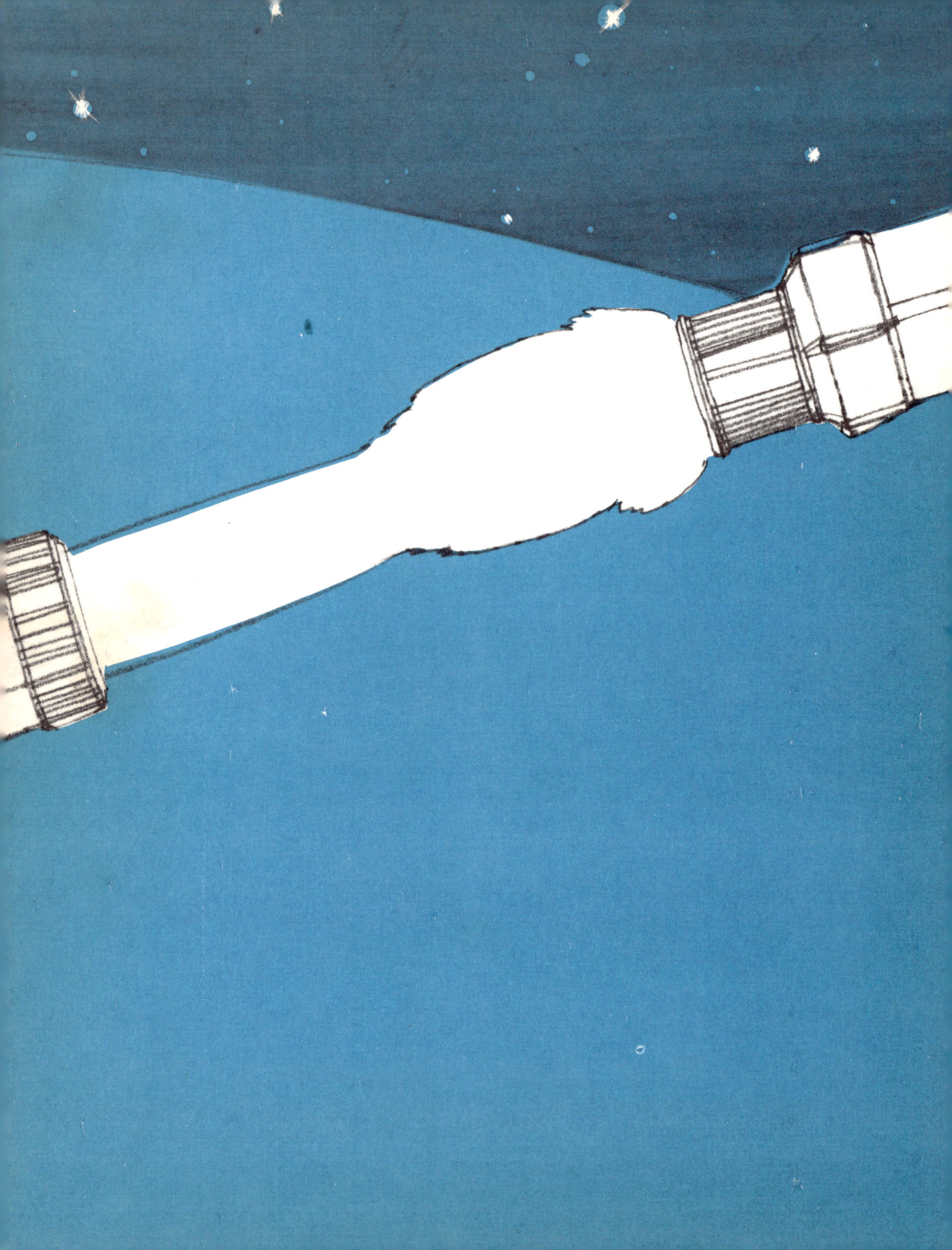

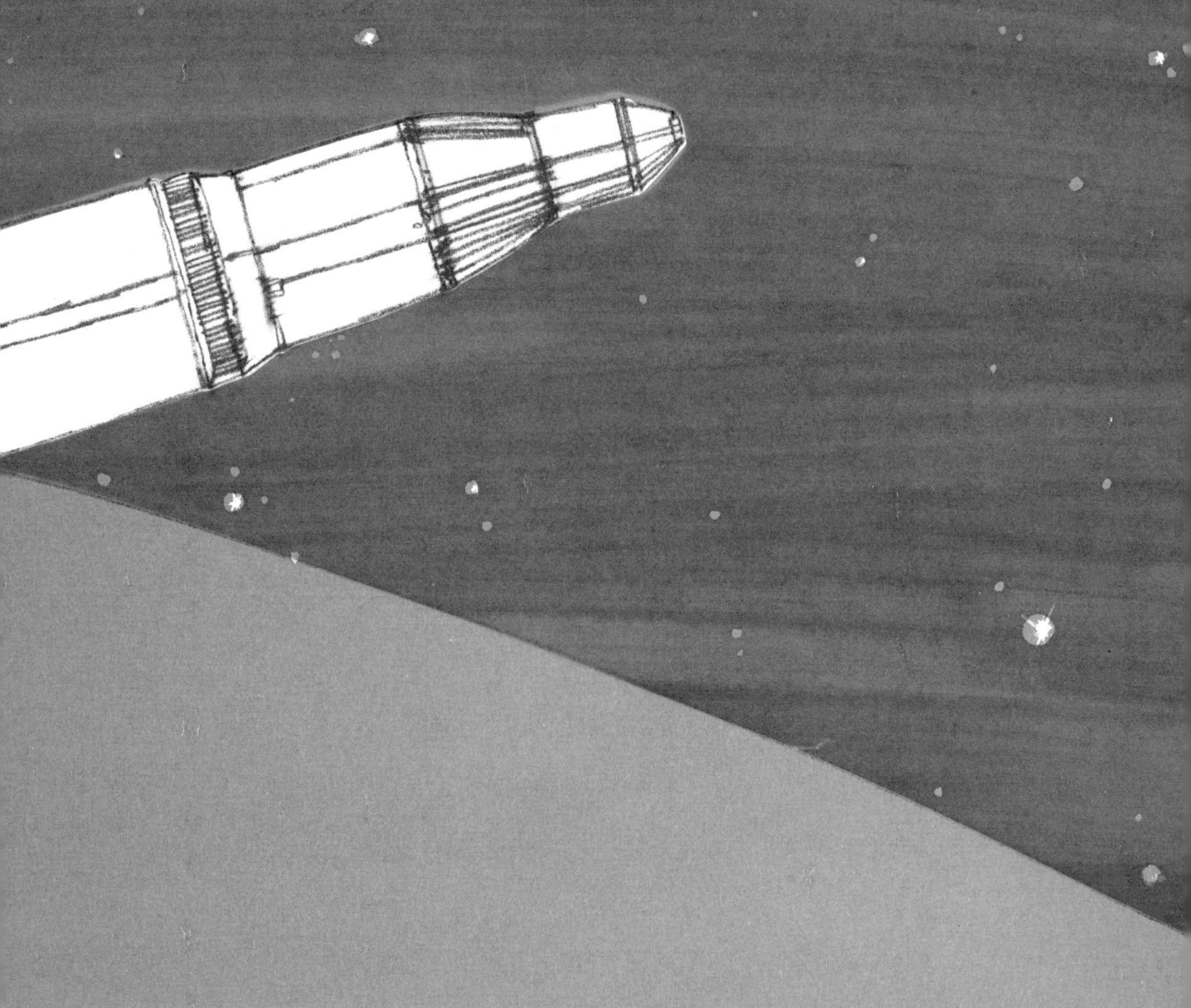

Faster and faster, higher and higher it rises.
The first rocket stage uses up its fuel and
    drops off.
Then the second stage burns out and drops.
The third stage fires, pushing the craft into orbit
    around the earth.

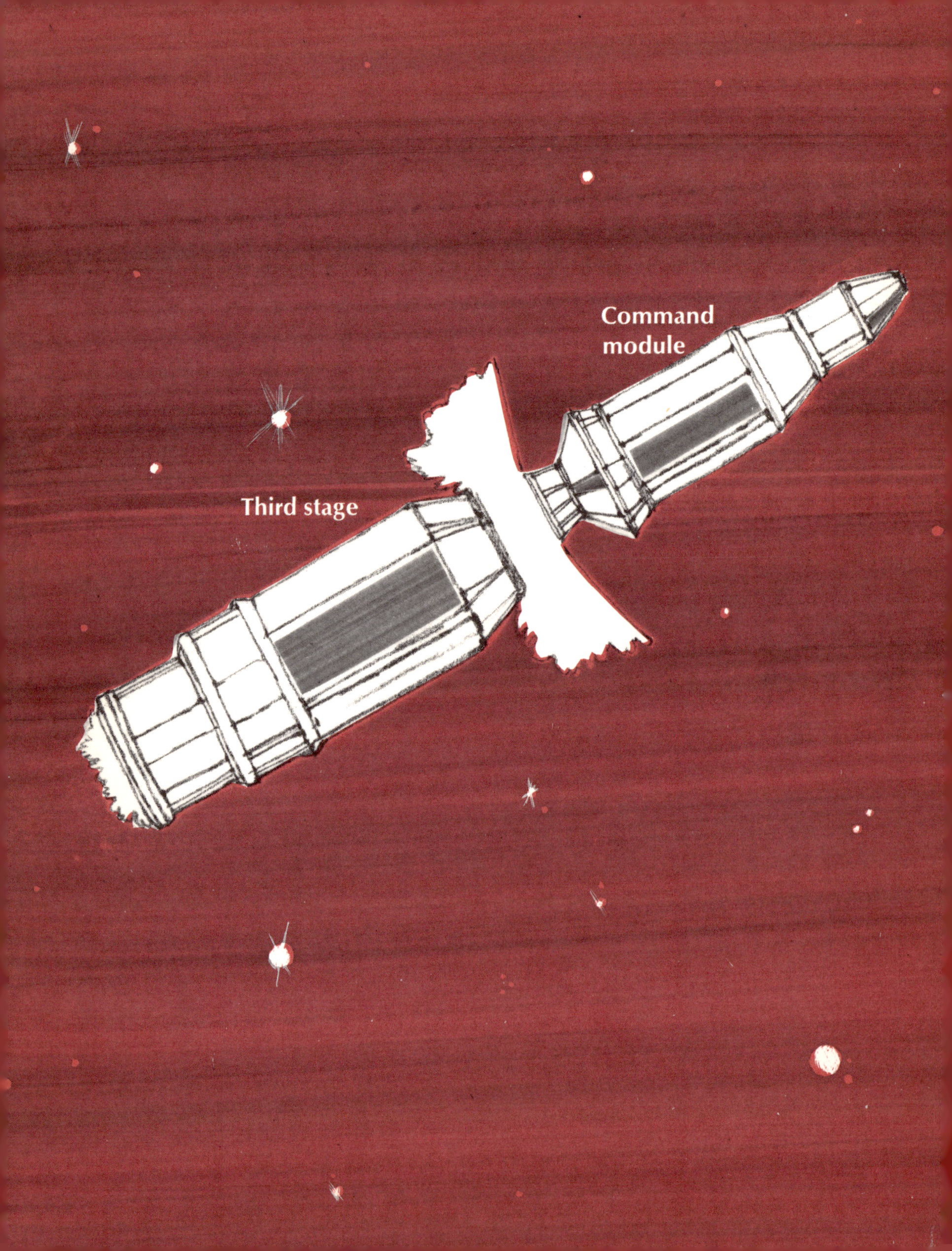

Command
module
Third stage

As it orbits the earth the craft is aimed at its
    target.
The third stage fires again, pushing the spacecraft
    out of earth orbit and toward the target.
Then the third stage drops off.
The spacecraft is shooting through space at
    speeds of up to 25,000 miles an hour.
Imagine! Over 7 miles a second!

In space flight, everybody and everything in the
capsule is weightless.

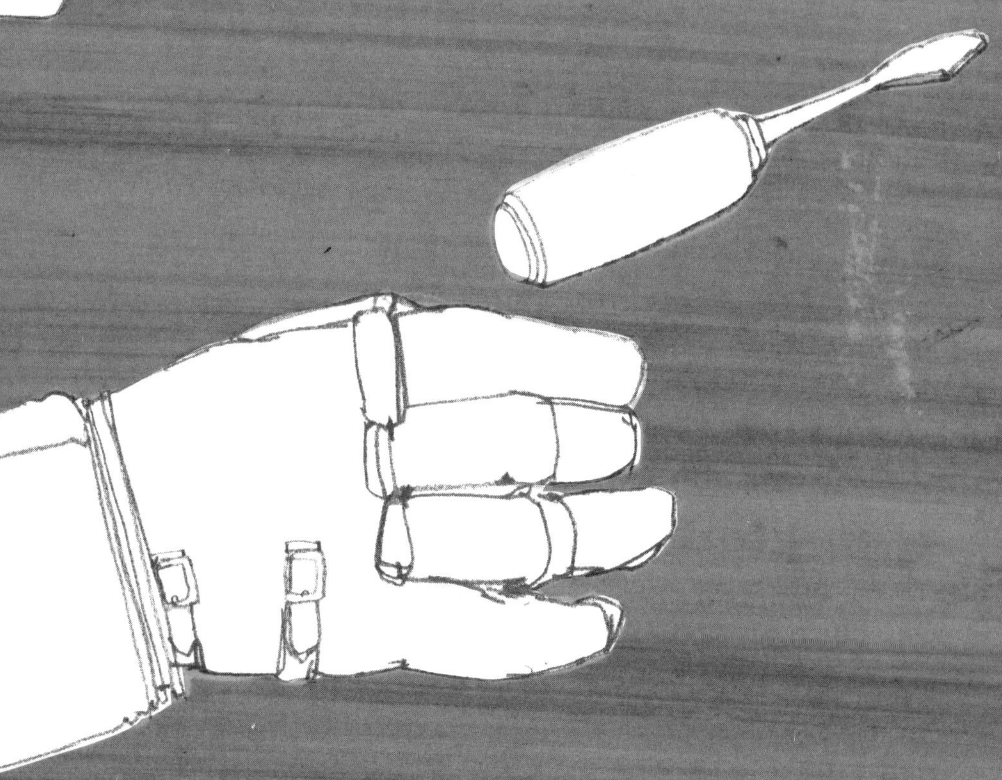

Nothing has weight.
Nothing can fall.
Let go of a tool and it floats in the air.
An astronaut in space flight can walk up a wall!

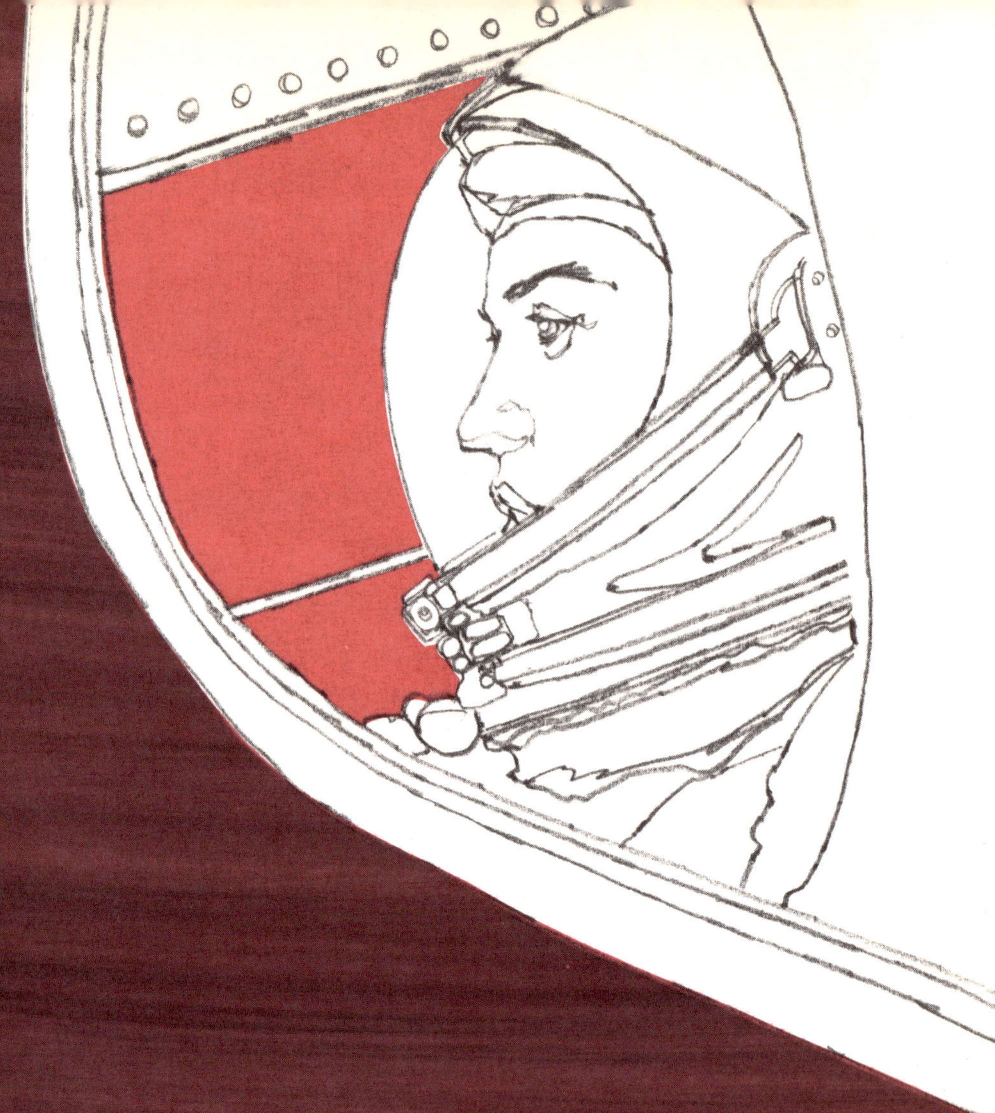

Space travel is dangerous.
Something may go wrong inside the spacecraft.
Or a tiny meteor can hit the capsule and put
  a hole in it.
If the hole is not fixed quickly, the air that the
  astronauts need will escape.

Space is filled with many dangerous, invisible rays. The walls of the command module must keep these rays out or the astronauts can be badly hurt.

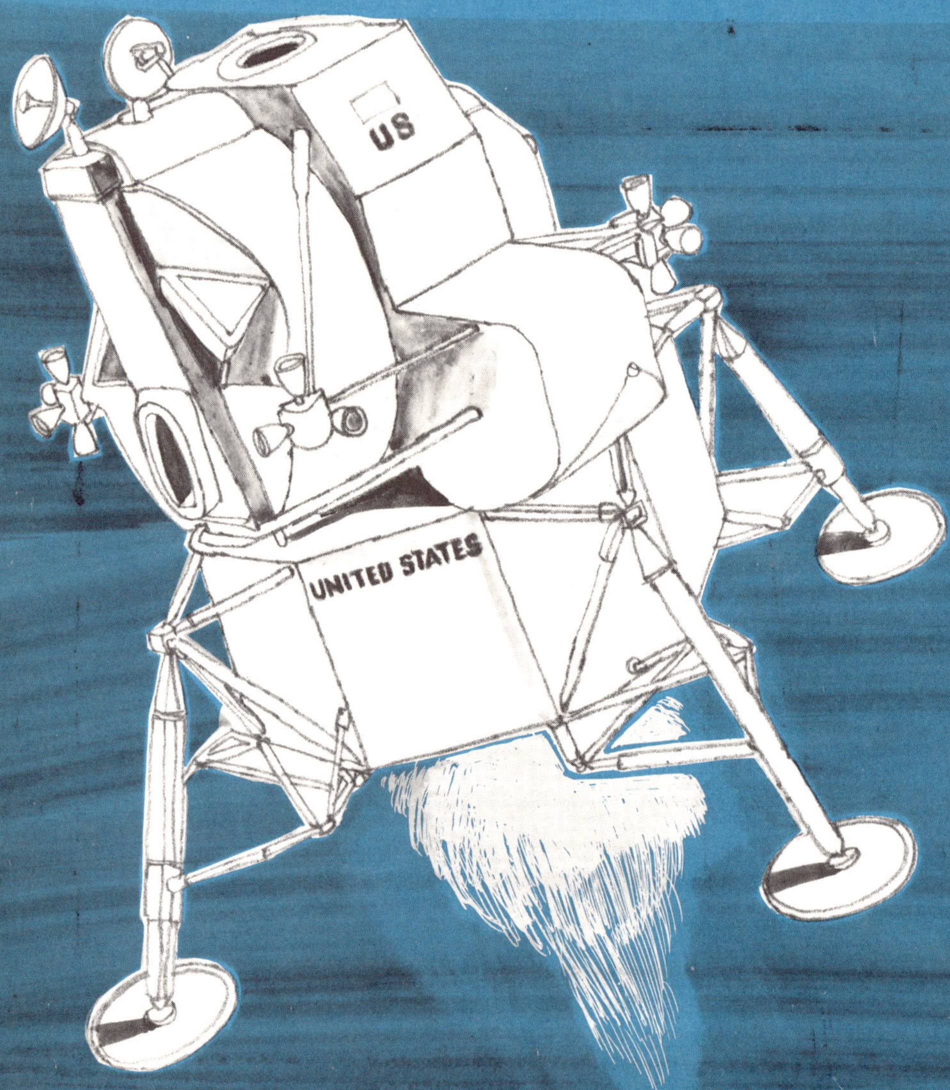

Landing module

The astronauts come down to the target in a
    landing module.
One astronaut stays in the command module.
He keeps it in orbit while the landing party—
    the men who land on the target—do their work.
Outside the landing module there is no air to
    breathe.
The men wear space suits while they work.
They have tanks full of oxygen to breathe.

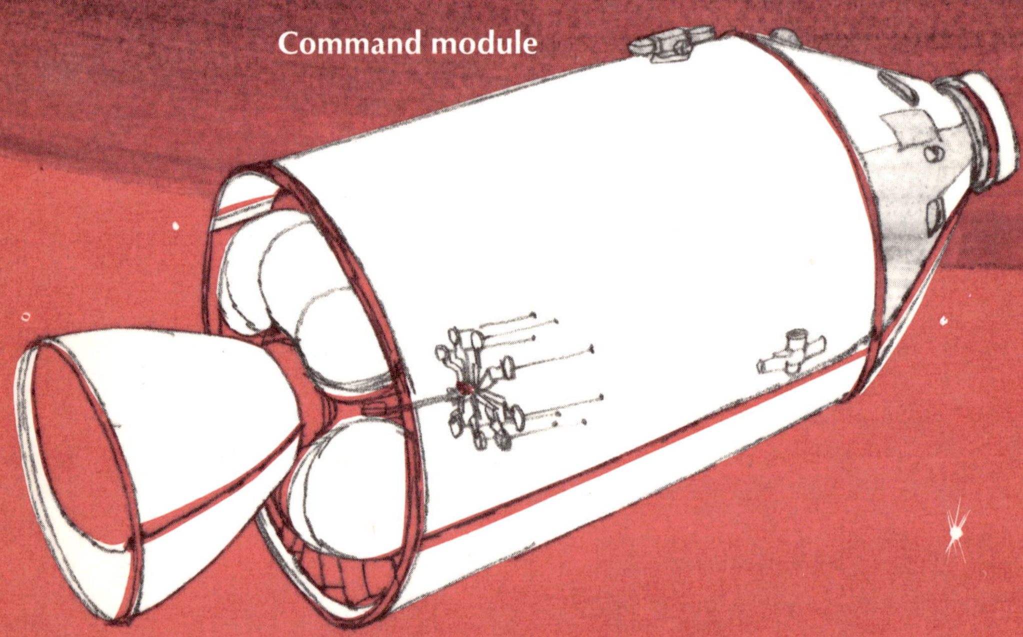
Command module

When the work is done, the landing party blasts
off in the landing module.
Back in orbit, the men dock with the command
module.

The landing party gets back into the capsule.
The landing module is let go, and the return
trip begins.

Rockets are fired to take the capsule out of orbit
around the target.
They start the capsule on its way back to earth.
As it hits the air layer around the earth the
capsule gets very, very hot.
The heat shield in front of the capsule gets
red-hot, but it protects the astronauts from
getting burned.

Parachutes open up to slow down the fall of the
capsule.
The capsule comes down to earth.
The astronauts are picked up and taken to space
headquarters.
There they tell scientists about their trip.

What will the next step in space travel be?
We may soon see space stations in orbit around
the earth.

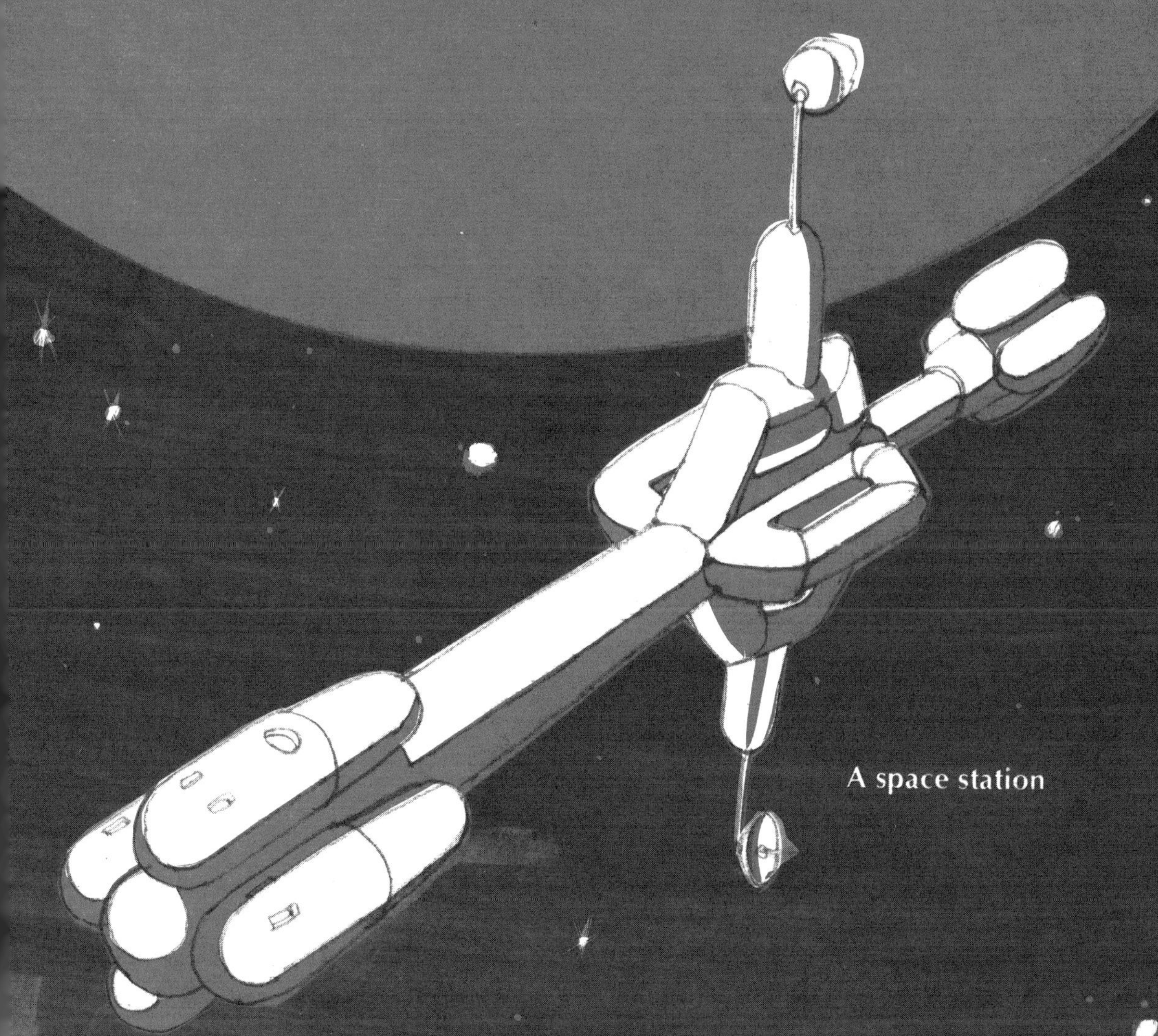

A space station

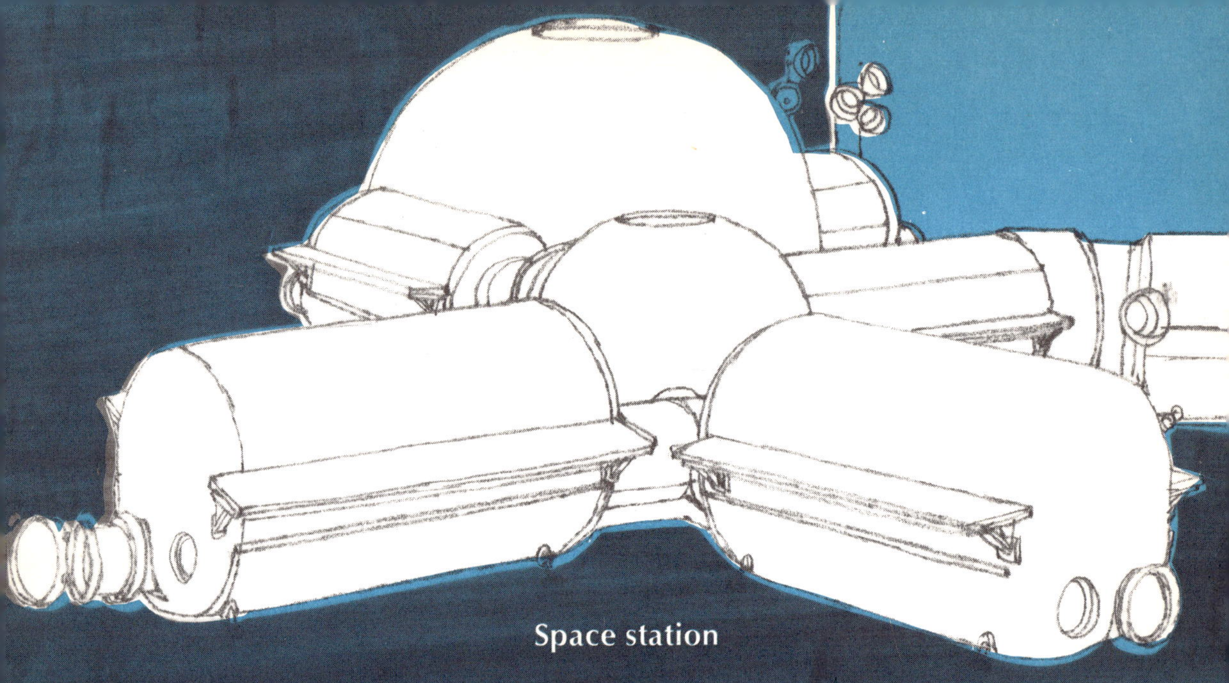

Space station

Or we may make the moon into a space station.
Spacecraft from earth will bring men and supplies
to the space stations.
Spaceships may be built on the space stations
and may then blast off to faraway planets.

Space ship

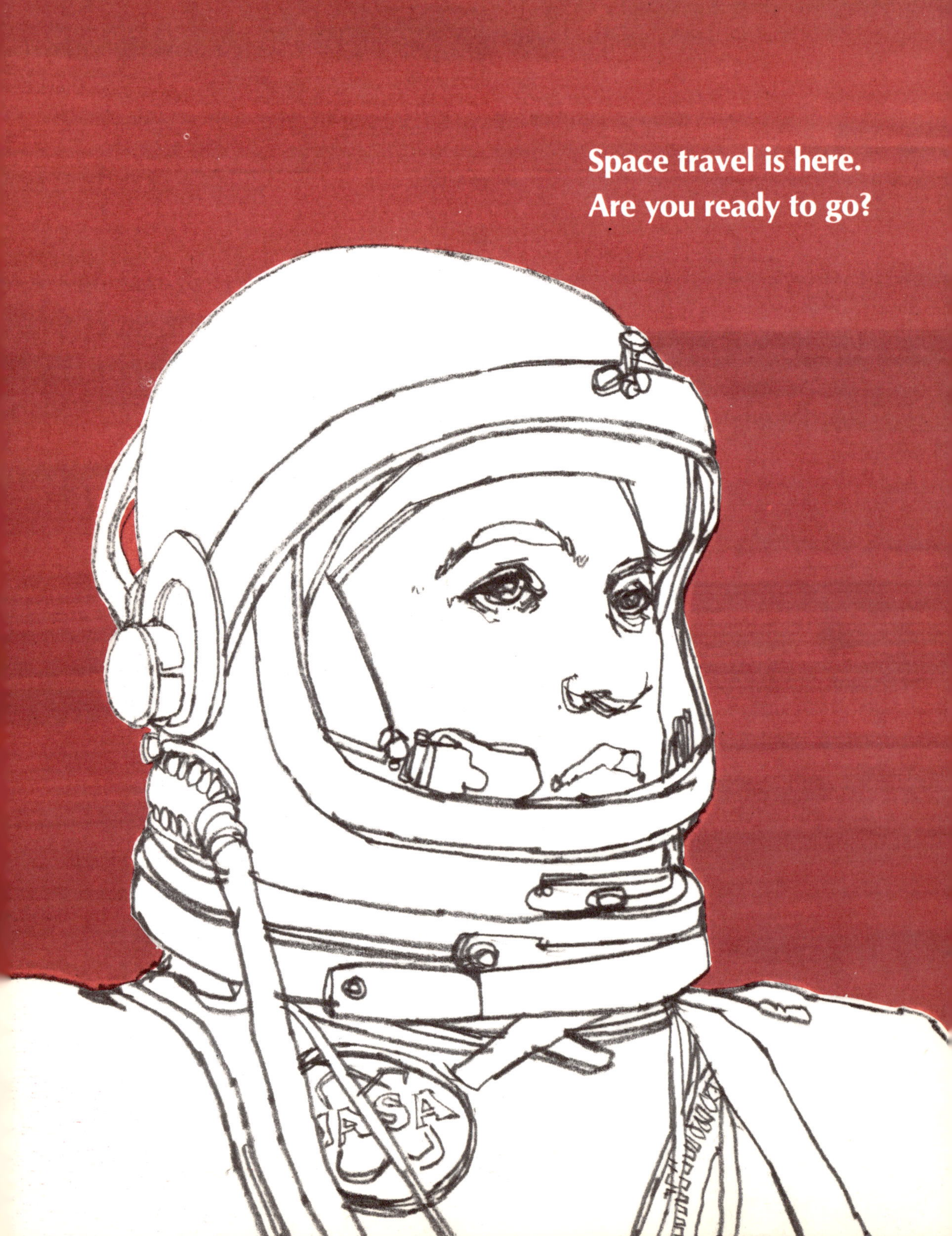
Space travel is here.
Are you ready to go?
NASA